NATIONAL GEOGRAPHIC

Ladders

WELCOME TO CHINA

AROUND THE WORLD

GENRE Social Studies Article

Read to find out about life in China's cities and villages.

WELCOME TO CHINA!

by Cynthia Clampitt

> Imagine trying to find a friend on this street! Shoppers clog the shopping district in Shanghai, China.

SAY IT IN CHINESE!

qíng (ching) means "please"

xìexìe (shyeh-shyeh) means "thank you"

Ní hăo mā? (NEE how mah) means "How are you?"

ON THE MOVE

Do you know how much one billion is? If you tried to count to one billion, it would take more than 30 years. The **population** of China, or the number of people living there, is nearly one and one-half billion. China has the largest population of any country in the world.

Where do the people of China live? They live in different kinds of communities. For many years, most Chinese people lived in **villages**. A village is a kind of community. It is smaller than a city. It is usually in the country. Many people work as farmers in Chinese villages.

Recently, many Chinese people have left villages. They have moved to big cities to work in factories or offices. Cities in China are very busy. Shanghai (shang-HI) and Beijing (bay-JING) are cities in China. They are two of the largest cities in the world. They are often crowded with people. All over China, people are always on the move!

Wó hén hǎo. (WUH hen how) means "I'm very well."

wǒ jìao (WUH jee-ow) means "my name is"

zài jìan (zai jee-en) means "good-bye"

A BUSY VILLAGE LIFE

Even in small Chinese villages, the people are very busy. Some people who live in villages work as teachers in schools. Some people who live in villages work as doctors in hospitals. Most villagers are farmers. Chinese farmers have to grow a lot of food to feed their country's population and to feed their own families.

Many Chinese farmers grow rice or wheat on their farms. Rice and wheat can be used to make many different foods. Even children help on farms. They take care of sheep and goats and other farm animals. They gather firewood so the family can build a fire to cook their meals. Children also pick fruit from fruit trees and sell it in local markets.

Children have time for fun after their farm work is done. They race each other to the market and float down rivers on rafts. Friends like to take bike rides together on country roads.

> Children who live in Chinese villages are very busy. They work on farms, do homework, and play with their friends.

^ In China, families like to eat together. This family eats food that has been grown on farms in their village.

^ Many village children help their families with farm work. They work before and after school.

^ Ducks are common farm animals in Chinese villages. Families eat their eggs. They sell the extras at the village market.

LIFE AMONG TALL BUILDINGS

Almost half of China's population still lives in villages. But Chinese cities are growing. Millions of people live in cities such as Shanghai and Beijing. Chinese cities are full of activity. During the day and during the night, people work, study, and have fun.

In most Chinese cities, families live in small apartments in tall buildings. More people fit in tall buildings than in short buildings. Tall buildings are important for cities with a lot of people. They make it easy for many people to live

Most city families don't have yards. Kids play in parks where there is plenty of space.

close together in a small amount of space.

The streets in Chinese cities are packed with people, bicycles, and cars. Trains and buses are very crowded. Bright signs flash. Buses and taxis honk. People rush across streets. Chinese cities are very busy places.

Trains in Chinese cities fill up quickly with many people.

These tall apartment buildings are in the Chinese city of Shanghai.

In Chinese cities, even bikes can get stuck in traffic jams.

Check In What do children do for fun in China?

GENRE Culture Article

Read to find out about the importance of rice in China.

Most rice paddies in China are built by hand. Some paddies have been used to grow rice for over 1,000 years.

THE UNDERWATER CROP

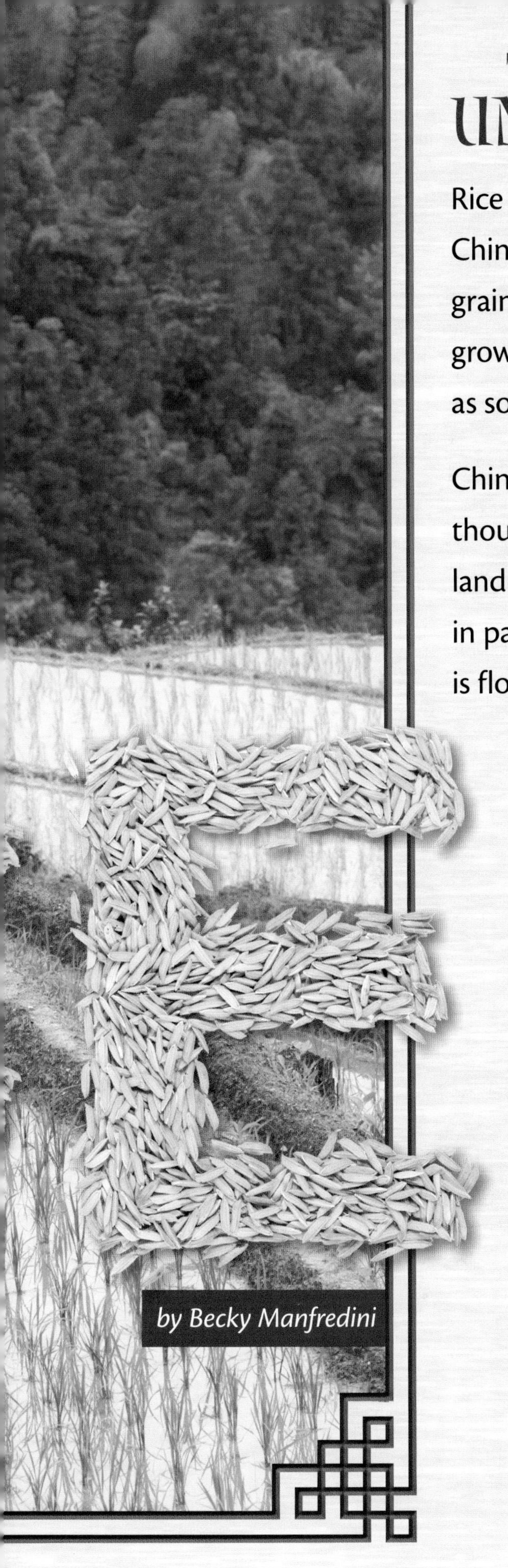

by Becky Manfredini

Rice is an important part of each meal in China. Rice is a small white or brown grain. It comes from a plant. Rice plants grow well in warm and wet places such as southern China.

Chinese farmers have grown rice for thousands of years. Farmers plow their land in the spring. Then they plant seeds in paddies. A **paddy** is an area of land that is flooded with about six inches of water. Many paddies look like stairs. Water from each step flows down to the next one. Rice grows well on flooded land. Most other crops do not.

Farmers drain the water from the paddies in the summer. They pull out any weeds they see. Then they flood the fields again so the rice can keep growing. They drain the paddies again in the fall. Then the farmers **harvest**, or gather, the rice crop. They let the rice dry. Then they take it to a building called a mill.

RICE FOR SALE

Each rice grain has a dry outer shell. The shell cannot be eaten. At the mill, workers and machines remove the shells. Then the rice is cleaned. This takes out rocks and dirt. Nobody wants to eat rice with rocks in it! After it is cleaned, the rice is safe to eat.

The clean rice is put in bags. The bags are put in boxes. The boxes are sent to the city. Then city families buy the rice in outdoor markets or in stores.

People sell rice in outdoor markets such as this one. Smart shoppers carry their rice home in a cart. Bags of rice are very heavy!

People have eaten rice for more than 5,000 years.

NOT JUST FOR EATING

Long ago, sticky rice was used as glue to hold walls and buildings together. Today, rice glue is used in many different ways, as shown below.

Need to write something down? For more than 1,000 years, Chinese people have been using rice to make paper. The surface of rice paper is smooth and white.

Chinese artists make little dolls out of rice powder and paper. People can buy these dolls in many of the outdoor markets.

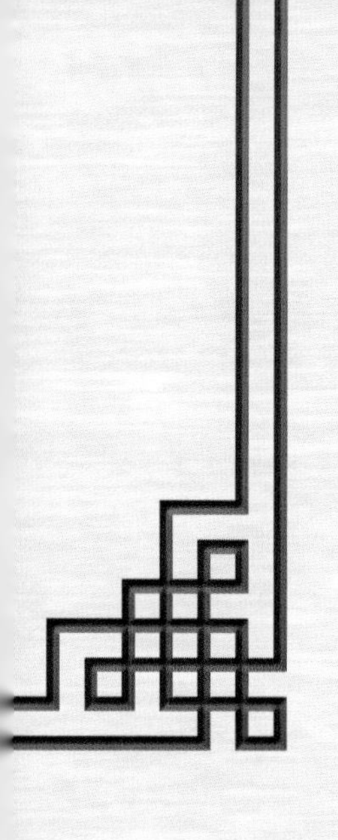

RICE FOR BREAKFAST, LUNCH, AND DINNER

The Chinese word *fan* means "rice." It also means "meal." Rice is served at every meal in China. You might think that the Chinese would get tired of eating so much rice. They don't. Eating rice in China is not boring.

Rice is served in many ways. For breakfast, you might have a bowl of rice porridge with cinnamon. Porridge is a soft

food made from rice cooked in milk. For lunch, you might have soup with rice. You might also have steamed sticky rice with vegetables. For dinner, you might have vegetables and chicken with rice. At every meal, you eat with **chopsticks** instead of a fork. Chopsticks are thin sticks used to pick up and eat food.

In China, rice can always be found on the dinner table. It's delicious by itself, or eaten with another dish.

HOW TO HOLD AND USE CHOPSTICKS

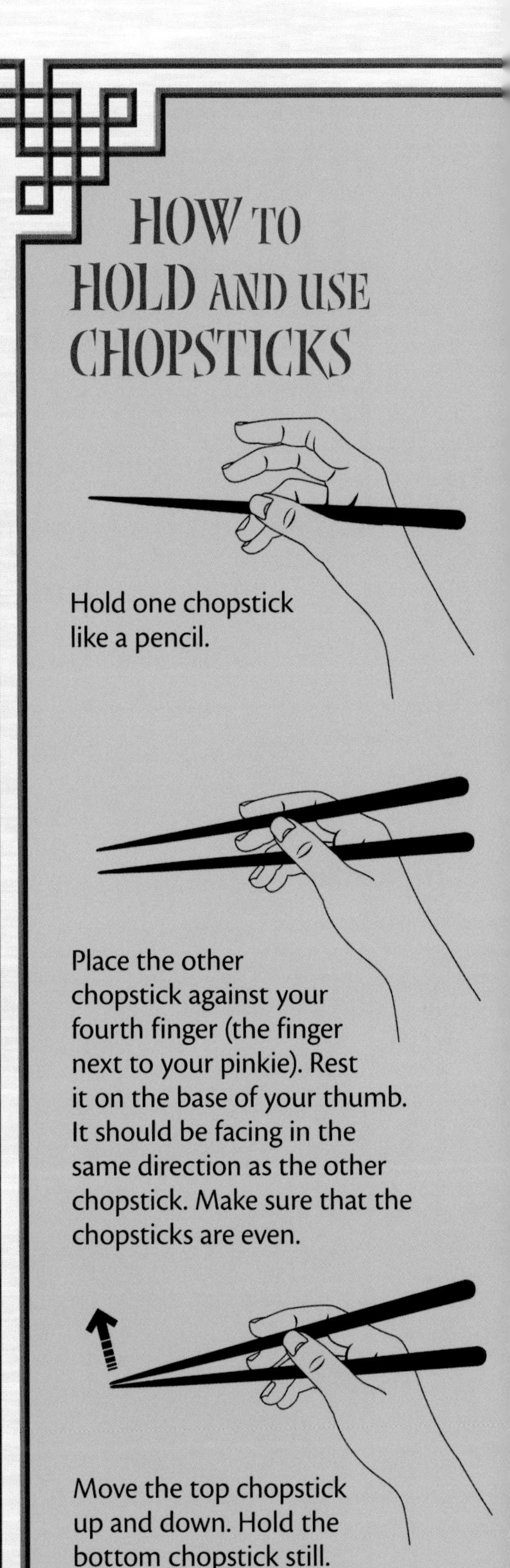

Hold one chopstick like a pencil.

Place the other chopstick against your fourth finger (the finger next to your pinkie). Rest it on the base of your thumb. It should be facing in the same direction as the other chopstick. Make sure that the chopsticks are even.

Move the top chopstick up and down. Hold the bottom chopstick still. Pinch a piece of stir-fried shrimp between the tips. Carefully bring the shrimp to your mouth. Mmm, mmm, good!

CELEBRATING WITH RICE

There are many holidays in China. On these fun days, Chinese people cook and eat foods made with rice.

Chinese New Year is a special holiday that takes place in January and February. It lasts for 15 days. People in China celebrate the new year by watching fireworks. They also dress in bright clothes. And they march in parades. Chinese New Year is a time when families share special meals. Many people make sticky rice cakes. They believe that these sweet cakes bring good luck in the new year.

The last event of Chinese New Year is the Lantern Festival. People hang lanterns in their homes and shops. Dancers perform with a big, colorful dragon. The dragon is a symbol of good luck. People make and eat sticky rice balls. The rice balls have nuts and fruits in them. The rice balls stand for family togetherness.

Rice is important to people all across China. In the country and in the city, this little grain plays a big part in Chinese people's lives.

Colorful dragon dancers carry a cloth dragon on poles. The lead dancer moves the head. The rest of the dancers move the long body.

Check In How do the people of China use rice in different ways?

GENRE Folk Tale

Read to find out about a popular Chinese folk tale.

The Thief and the Elephant

retold by Elizabeth Massie
illustrated by Rich Lo

Stories are an important part of a country's culture. They entertain us. They can also teach us important lessons about life. A folk tale is a story passed down by families for many years. This Chinese folk tale teaches that honest people do not need to fear the truth.

Long ago in China, elephants were taught to do many things. They helped farmers plow the land. They helped build roads. They even helped woodcutters carry heavy logs. People knew elephants were strong. They also believed elephants were wise. Many people thought an elephant could tell if a person was lying or telling the truth.

One day, a woman went to the court. She was very upset. "I was robbed!" she told the judge. "A man came into my house last night. He stole my jewelry. I want him to be found. He should be punished." She said that she saw the man who robbed her. She told the judge what the man looked like.

The judge told her he would solve the crime. He sent his officers to find five men who looked like the robber. The five men were brought to court.

All five men were about the same height. They had the same haircuts. When the woman came into the courtroom, she looked at each man. She pointed to one man and said he was the thief.

"Are you sure?" asked the judge.

The woman said, "Yes. That is the man!"

Then the judge opened a large door. He led an elephant into the court. "This is my elephant. She can tell what a person is thinking and feeling. She will know who robbed you."

Four of the men smiled as the elephant touched their heads and chests with her trunk. The fifth man trembled in fear. His face became red with worry. He was not the man the woman said had robbed her.

The judge watched the elephant. Then he shouted, "Do your duty!" The elephant wrapped her trunk around the trembling man and lifted him. She set him down in front of the officers. The frightened man said that he was the thief. The officers took him to jail.

The woman was surprised. "I was sure I was right," she said.

The judge told her, "You must always be careful whom you accuse."

The woman nodded her head and replied, "I will, sir."

The judge patted his elephant and said, "You have helped me. There is good food in your stable. Go now and eat it."

The elephant went back out through the large door. She swung her mighty trunk and waved her huge ears. The woman bowed to the judge and left.

The next day, three men came to the courtroom to talk to the judge. They had many questions for him.

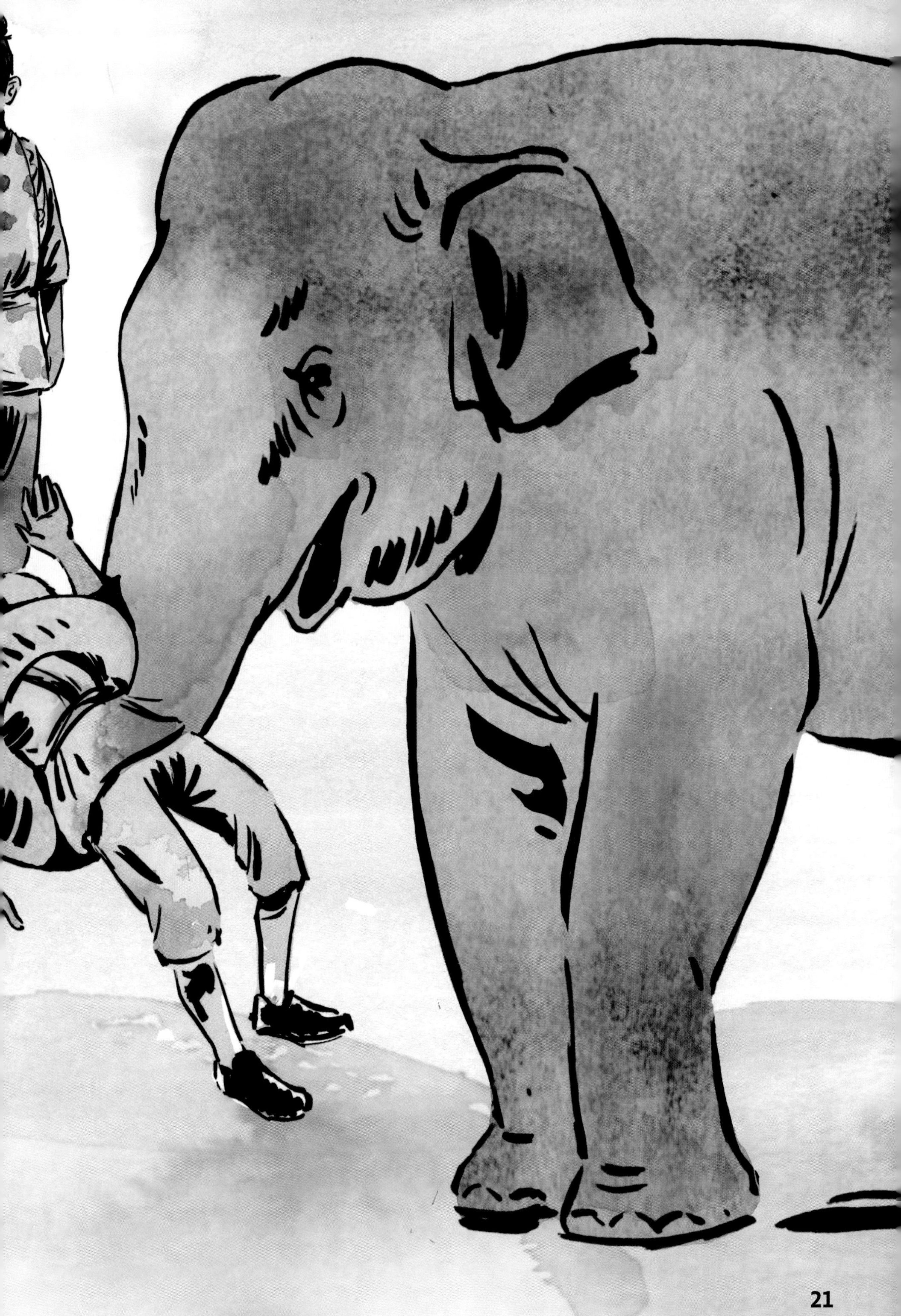

“We heard about your elephant,” said the first man. “We are amazed. It is hard to believe that an animal can know what a man is thinking and feeling.”

“Yes,” agreed the second man, “she picked the thief out. How did your elephant get this power? Does she eat special food? Did you teach her tricks?”

The judge laughed. He said, “My elephant eats what all elephants eat. I have not taught her any tricks.”

The men were confused. The judge told them, “Many people believe elephants can tell if a person has something to hide. The five men who were here yesterday believed it. The four honest men knew they had done nothing wrong. They were not worried. But the thief was afraid. He believed the elephant would discover his crime.”

The third man said, “He could not hide his fear.”

“Yes,” said the judge. “His fear showed us that he was guilty.”

The three men left the courtroom.
They were grateful to be honest men.
They knew that life is much easier
when you have no fear.

Check In Why was the guilty man scared of the elephant?

Discuss

1. What do you think connects the three selections that you read in this book? What makes you think that?
2. How is the life of a child who lives in a Chinese village different from that of a child who lives in the city?
3. How is rice an important part of Chinese celebrations?
4. Do you think you would have been able to pick out the guilty man in the folk tale? Why or why not?
5. What do you still wonder about life in China? How can you learn more?